SpringerBriefs in Applied Sciences and Technology

SpringerBriefs present concise summaries of cutting-edge research and practical applications across a wide spectrum of fields. Featuring compact volumes of 50 to 125 pages, the series covers a range of content from professional to academic.

Typical publications can be:

- A timely report of state-of-the art methods
- An introduction to or a manual for the application of mathematical or computer techniques
- A bridge between new research results, as published in journal articles
- A snapshot of a hot or emerging topic
- An in-depth case study
- A presentation of core concepts that students must understand in order to make independent contributions

SpringerBriefs are characterized by fast, global electronic dissemination, standard publishing contracts, standardized manuscript preparation and formatting guidelines, and expedited production schedules.

On the one hand, **SpringerBriefs in Applied Sciences and Technology** are devoted to the publication of fundamentals and applications within the different classical engineering disciplines as well as in interdisciplinary fields that recently emerged between these areas. On the other hand, as the boundary separating fundamental research and applied technology is more and more dissolving, this series is particularly open to trans-disciplinary topics between fundamental science and engineering.

Indexed by EI-Compendex, SCOPUS and Springerlink.

Eduardo Sánchez-García · Javier Martínez-Falcó ·
Luis A. Millán-Tudela · Bartolomé Marco-Lajara

Sustainable Management Through Knowledge and Innovation

How to Develop a Strong Strategy in the Wine Industry

 Springer

Eduardo Sánchez-García
Department of Management
University of Alicante
San Vicente del Raspeig, Alicante, Spain

Luis A. Millán-Tudela
Department of Management
University of Alicante
San Vicente del Raspeig, Alicante, Spain

Javier Martínez-Falcó
Department of Management
University of Alicante
San Vicente del Raspeig, Alicante, Spain

Bartolomé Marco-Lajara
Department of Management
University of Alicante
San Vicente del Raspeig, Alicante, Spain

ISSN 2191-530X   ISSN 2191-5318 (electronic)
SpringerBriefs in Applied Sciences and Technology
ISBN 978-3-031-64791-8   ISBN 978-3-031-64792-5 (eBook)
https://doi.org/10.1007/978-3-031-64792-5

This Springer imprint is published by the registered company Springer Nature Switzerland AG
The registered company address is: Gewerbestrasse 11, 6330 Cham, Switzerland

If disposing of this product, please recycle the paper.

Preface

Wine has been produced since immemorial times. In fact, there are records as old as Mesopotamia, which gives us an idea of the scope of this activity.

As it was at prior times, the production and selling of wine is not simply an economic issue nowadays, since it has always been linked to the history of mankind. From its veneration to its prohibition, numerous civilizations have found both a nutritional and recreational activity in this beverage. That is why we can consider wine as an element linked not only to the economy in various historical periods, but also to the socialization, culture and even technology of each of these stages that have shaped the world as it is today.

Indeed, traditional economic activities have made the world what it is today. It is undeniable that the changes brought about by the two Industrial Revolutions occurred in response to a need for more and better production of goods. Nor can we hide the fact that these changes have carried out a transformation in all economic sectors. In short, there is a symbiosis between economic activity and the society to which it relates. This question, which may seem obvious, entails a great and important implication: economic activities must be congruent with the reality in which they find themselves, which irremediably requires an alignment of both. This is precisely what has motivated us to write this work.

The wine activity in all its extension (from its historical legacy to the latest techniques developed) is widely spread throughout the world, being the livelihood of a large number of families and individuals. Taking into account this reality, as well as the demands of today's society toward a sustainable, responsible and consistent economy in all its dimensions, we have prepared this brief work so that those who wish to venture into the field of sustainability in the wine industry can acquire the basic notions that will allow them to start on this path. That is why we, the authors, would like to emphasize that this book is nothing like a manual of good practices or a protocol for the sake of sustainability. Our purpose is that those people who wish to understand what sustainability implies in the wine industry have a tool that brings it closer to human language, regardless of whether the reader has extensive knowledge about the latest developments in the field of wine or is a person linked to the traditional part of the sector.

In this sense, and with the aim of offering condensed information, in the following chapters we will make a first incursion into the management of sustainability in the wine industry, introducing the basic concepts necessary to understand the current lines of action in this respect, as well as to explain some of the challenges and opportunities that arise from the implementation of sustainable measures in the wine sector. We continue with a series of theoretical questions about sustainability, so that the reader can acquire a series of essential notions with which to interpret and understand the most elementary concepts of sustainability, as well as innovation as a way for its implementation. Once this phase is completed, we will focus on how to develop business strategies that incorporate sustainability, analyzing its strategic value and the need to involve the various stakeholders present in the sector. As the last contents, we will work on the concept of learning organization as a frame of reference on which to transfer all the above to reality, taking this model as the vision of a wine company that effectively incorporates the three dimensions of sustainability (economic, social and environmental) while simultaneously considering the external conditions it must face. Finally, we close this work with some brief conclusions that condense all of the above, so that the readers have a narrative compilation on which to reflect and from which to draw their own conclusions, considering the one-person reality that they must face in this regard.

Without further ado, we hope that you find the content of this work useful and interesting, as was our intention in writing each and every line of this work.

San Vicente del Raspeig, Spain
Eduardo Sánchez-García
Javier Martínez-Falcó
Luis A. Millán-Tudela
Bartolomé Marco-Lajara

Contents

Chapter 1
Introduction to Sustainable Management in the Wine Industry

1.1 Overview of the Wine Industry

Sustainable management in the sector of winemaking, despite being a relatively recent development, is deeply linked to the tradition of the sector, that has centuries of history (Da Silva Lopes et al. 2020), a tradition that has seen the art of winemaking evolve in response to the changing dynamics of our world, and has shaped the current practices of wine production, strengthening the relationship between humanity and wine, a relationship that has been both a source of artistic expression and a reflection of the environmental and societal contexts of each era (Forbes et al. 2020).

The origins of winemaking date back to 6000 BCE in Mesopotamia and Egypt. By that time, wine was already recognized as a symbol of cultural and religious significance, highly associated with deities and, consequently, used in ceremonial practices (Tiefenbacher and Townsend 2020). Greeks and Romans elevated the prior status, integrating wine in other facets of the daily life, irrigating societal events as a sign of sophistication and prosperity. Indeed, the vineyards farmed by Greeks and Romans are the direct forefathers of the ones we have nowadays, something that also applies to growing techniques (Anderson and Pinilla 2022).

As the Roman Empire expanded, so did the grape-growing and wine production in the new territories. This led to the generation of the diversity of wine cultures that we have nowadays. However, the fall of the Roman Empire, together with the appearance of the Middle Ages, led to winemaking move from a democratized societal scope to its focus on the monastic communities, who played a crucial role in preserving and refining viticulture techniques during this period (Ayuda et al. 2020). Monasteries developed a methodical approach that is currently known by the recording of the experimentation they did, so they can be considered as early centers of viticultural knowledge and innovation (Villanueva and Ferro 2022).

Moving to the Renaissance, which was a period of renewed interest in arts and science, viticulture showed a similar behavior, being it materialized by the significant advancements in the knowledge of grape cultivation and wine production, putting

E. Sánchez-García et al., *Sustainable Management Through Knowledge and Innovation*, SpringerBriefs in Applied Sciences and Technology, https://doi.org/10.1007/978-3-031-64792-5_1

more emphasis on quality and terroir and, thus, recognizing the importance of the environment in the final product. During this time, the concept of appellation (a designation of origin to denote the quality and character of wines that still remains in use) began to take shape (Charters et al. 2022).

The Industrial Revolutions that took place in the eighteenth and nineteenth centuries also showed to be turning points in the history of wine. As it could be expected, the development of specialized machinery, together with considerable improvements in means of transportation, brought a transformation in the wine industry, both in terms of production and distribution (Millon 2013). Winemaking, an activity that had been local and artisanal up to that point, evolved into a more global, industrialized activity, with important transformations in bottling and storage, as well as the possibility to large-scale operation, thus being able to meet the growing demand around the world (Anderson and Pinilla 2018).

The twentieth century showed a similar trend to the prior stage: fast growth and industrialization. However, they were accompanied by important environmental challenges, such as the widespread use of chemical fertilizers and pesticides, overexploitation of water resources and loss of biodiversity in wine-growing regions. All these elements generated a concern about the long-term viability of the new viticulture practices (Morrison and Rabellotti 2017). In response to those challenges, the concept of sustainable management in the wine industry began to emerge, returning to some earlier practices that were more environmentally friendly. However, this was not at the expense of the latest scientific and technological advancements, as Campbell and Guibert (2006) highlight.

As we can see, the evolution of the wine sector is as rich and complex as the beverage itself. From the ancient Mesopotamian vineyards to the modern wineries, the history of wine is one of ingenuity and creativity, of adaptation and innovation. If we focus on the twenty-first century, that evolution has translated into the principles of sustainable management. These are furtherly developed through this chapter. The wine industry is helping to preserve the delicate balance of our ecosystem while ensuring that the art of winemaking, deeply rooted with the traditions of the sector, continues to evolve.

1.2 Environmental, Economic and Social Aspects

The concept of sustainability encompasses a broad spectrum of interlinked aspects, which are specific to what is being assessed. All those aspects can be sorted into three basic dimensions: environmental, economic and social, being winemaking no exception (Flores 2018).

Focusing on the wine industry, and starting with the environmental aspects, the relationship between vineyards and their surrounding ecosystem stand out, as the equilibrium between both parts is delicate and sensitive. In this group, we can include sustainable soil management practices (reduced tillage, cropping covering, or the use or organic compost) to maintain its fertility or the preservation of biodiversity in the ecosystems where the vines are grown (for example, by preserving autochthonous varieties which also help the preservation of fauna and flora). A problem that emerges from viticulture is the use of water resources, as vineyards can be water-intensive (particularly in arid and semi-arid regions), so optimization of this resource becomes a critical aspect regarding environmental sustainability, as exposed by Mariani and Vastola (2015). Practices aligned with this goal are drip irrigation systems, rainwater harvesting and the recycling of wastewater (being some more difficult than others). These operations are not only intended for short-term water supply, but to ensure the long-term availability and quality of water resources in the areas where they are applied (Costa et al. 2022).

The aforementioned biodiversity preservation is another goal that can be achieved by the winemaking sector, as vineyards can become (if properly managed) hotspots for biodiversity, as they can act as support for a wide range of plant and animal species and, consequently, contribute to the ecological health of the region (Christ and Burritt 2013). This preservation can be achieved by means of planting cover crops, reducing the use of chemical pesticides and herbicides, as well as ensuring that the introduction of new grape varieties does not affect the autochthonous species.

Regarding economic sustainability, we must highlight that the wine industry can help ensuring the economic viability and profitability of investments in the area. One of the key economic aspects of sustainability is the concept of sustainable yield, which is about producing the optimal quantity of grapes that can be sustainably harvested from a vineyard, allowing to find the spot where the quantity of grapes produced is in harmony with the capacity of the vineyard to sustain that production over the long term, so that sustainable yield practices, such as careful vineyard management and yield regulation, help to ensure that the vineyards remain productive and profitable (Dodds et al. 2013).

Furthermore, nowadays sustainable wines, and particularly those that are certified as organic, biodynamic, or sustainably grown, offer a point of differentiation that can be a key factor in attracting and retaining customers, so that market differentiation is another important economic aspect of sustainability in wine production, being about tapping into a niche market, as well as responding to a growing consumer demand for products that are of high quality as well as environmentally and socially responsible (Nave et al. 2021). This demand is driven by a growing concern among consumers about the impacts of the products they consume have, so this sort of certifications has become an opportunity for product differentiation (Signori et al. 2017). Another point in favor of economic sustainability coming from this sector is wine tourism. If properly developed (i.e., if it is aligned with environmental concerns) can enhance rentability of vineyards, as its development is based mainly in intangible assets.

While the initial investment in technologies such as renewable energy systems or water-efficient irrigators can be significant, they may lead to long-term cost savings and improved profitability, if properly designed. If they are combined with other investments in research and development (R&D), all of them together can make the sector a competitive one in the increasingly sustainability-conscious market, as pointed out by Pratt (2012).

Ending with social sustainability, the wine industry is related to it by different means. We can find that people who make wine production possible and communities related to them depend not only economically and environmentally. Their culture, traditions and history are usually rooted to the sector. Wineries, particularly those in rural and regional areas, are often key players in their local communities, providing employment, economic activity and a sense of identity and pride, so that sustainable community engagement practices, such as supporting local businesses, participating in community events and contributing to local environmental and social initiatives, help to build strong and resilient communities, giving back or being a good corporate citizen and recognizing that the success and sustainability of the wine industry are inextricably linked to the health and well-being of the communities in which it operates (Touzard et al. 2016).

Wine is a beverage that can be assessed as a cultural asset, as it is a reflection of the history, traditions and values of their origin regions, thus contributing to social sustainability. If traditional winemaking practices are developed, as well as the preservation of local varietals and practices addressed toward achieving environmental sustainability, we already have two pillars of sustainability already covered. Add its economic contribution, and we obtain the perfect cocktail for sustainability. Therefore, the importance of sustainability in the wine industry, encompassing environmental, economic and social aspects, is a comprehensive and holistic approach that is essential for the long-term viability and integrity of the industry, so that as the wine industry continues to evolve and respond to the challenges and opportunities of the twenty-first century.

1.3   Unique Challenges and Opportunities in Implementing Sustainable Practices in Viticulture and Enology

The path toward sustainable management is conditioned by different challenges and opportunities that exert their influence simultaneously, each of them presenting their own difficulties and prospects for innovation, which are not isolated phenomena. Indeed, they are deeply interconnected, each influencing and shaping the other in a dynamic interplay that reflects the always-changing nature of the wine industry (Flores and Medeiros 2016).

One of the main challenges in the pursuit of sustainability by the wine industry is the adaptation and mitigation of climate change, as it has a remarkable impact on viticulture (as with many other flora), due to phenomena such as rising temperatures, changing precipitation patterns and increased frequency of other extreme weather events (Luzzani et al. 2021). In this regard, the transition to viticulture and enology practices more aligned with sustainability come together with major investment costs than become a strategic barrier for their implementation, as Mariani and Vastola (2015) pose. However, they are also an opportunity (if those barriers can be avoided) as consumer demand for sustainable wines is increasing, so they can become differentiated products with consequent premium prices, thus contributing to the economic resilience and profitability of sustainable wineries in the long term (Figueroa and Rotarou 2018).

One opportunity for the wine industry to improve its sustainability is the fact that it is deeply rooted in tradition, with centuries-old practices and varietals that, while being linked to the cultural identity of the wine-producing regions, are suitable for exploitation, and even desirable in cases where differentiation can play a remarkable role. Indeed, the application of those practices while simultaneously using autochthonous varietals is a way for the creation of unique value propositions (Corbo et al. 2014).

In this vein, the efficient and responsible use of water is an important concern in many wine-producing regions, particularly those facing water scarcity and drought conditions, being also the preservation of soil health and fertility essential for the long-term productivity and sustainability of vineyards (Ollat et al. 2016). These challenges require a multi-dimensional approach to resource management, considering the interdependencies among the different ecosystem elements.

Regarding to the social dimension of sustainability, building a more equitable and inclusive industry that values and supports its workers and communities, ensuring fair labor practices, worker safety and community well-being is essential, since initiatives such as fair-trade certification, community development programs and employee training and development can enhance the social sustainability of the industry, creating a more committed and motivated workforce and a stronger connection with local communities (Montalvo-Falcón et al. 2023).

Moreover, the integration of sustainable practices across the entire value chain of wine production, from the cultivation of grapes to the bottling and distribution of wine, presents a unique set of environmental, economic and social challenges, e.g., reducing the carbon footprint in the transportation and distribution of wine requires logistical innovations and collaboration with partners across the supply chain, as well as an opportunity to revolutionize the industry through the adoption of green logistics, sustainable packaging solutions and efficient distribution networks, so that by rethinking and redesigning the supply chain, the wine industry can significantly reduce its environmental impact and set new standards in sustainability (Goncharuk 2017).

The process of winemaking generates a substantial amount of waste, including grape skins, stems and seeds, which if not managed properly can lead to environmental pollution, so that innovative practices such as composting grape waste to enrich vineyard soils, extracting valuable compounds from grape seeds and skins for use in cosmetics and health products and utilizing wastewater for irrigation are examples of how the industry can turn waste into a resource, thereby reducing its environmental footprint and creating additional streams of revenue, being these practices integrated in what is known as circular economy.

Monoculture practices, commonly employed in conventional viticulture, can lead to a loss of biodiversity and an increased reliance on chemical inputs, so that practices such as intercropping, maintaining natural habitats and fostering beneficial insects and wildlife enhance biodiversity and help to the contribution of the vineyard natural resilience, reducing the need for chemical inputs and enhancing the overall health of the ecosystem (Christ and Burritt 2013).

Despite the aforementioned increasing interest in sustainable wine, a huge base of consumers is still unaware of the nuances of sustainable viticulture and the sustainable impacts (related to its three dimensions) of their wine choices, as indicated by Lichy et al. (2023). Thus, education and transparency in this regard is a must. By increasing the consumer knowledge on the topic would increase the demand for sustainable wines, creating a virtuous cycle that supports the growth and development of sustainable practices in the industry. Moreover, the wine industry (as part of the alcoholic drinks sector) is subject to specific regulations that vary by regions, being sometimes a barrier for profitability and, thus, generating an impediment for the investment of surpluses in sustainable practices. By collaborating with different stakeholders, the wine industry may help to shape policies that promote sustainability, with the potential of extrapolating them to other agriculture-related sectors (Gomes et al. 2021).

Summing up, all these challenges are also an opportunity for differentiation via innovation, transformation and leadership. The potential growth in these techniques, if they can be deployed across the industry, would lead it to a more sustainable and resilient future, through balancing environmental stewardship, economic viability and social responsibility (Angelova et al. 2019).

References

K. Anderson, V. Pinilla, *Wine globalization: A new comparative history* (Cambridge University Press, 2018)

K. Anderson, V. Pinilla, Wine's belated globalization, 1845–2025. Appl. Econ. Perspect. Policy **44**(2), 742–765 (2022)

M. Angelova, D. Pastarmadzhieva, G. Dimitrova P. Georgiev, Innovative practices in the wine industry: Opportunities for competitiveness enhancement in Bulgaria, in *2019 International Conference on Creative Business for Smart and Sustainable Growth (CREBUS)* (IEEE, 2019), pp. 1–6

M.I. Ayuda, H. Ferrer-Pérez, V. Pinilla, Explaining world wine exports in the first wave of globalization, 1848–1938. J. Wine Econ. **15**(3), 263–283 (2020)

G. Campbell, N. Guibert, Introduction: Old World strategies against New World competition in a globalizing wine industry. Br. Food J. **108**(4), 233–242 (2006)

S. Charters, M. Demossier, J. Dutton, G. Harding, J.S. Maguire, D. Marks, T. Unwin (eds.), *The Routledge handbook of wine and culture* (Routledge, 2022)

K.L. Christ, R. Burritt, Critical environmental concerns in wine production: An integrative review. J. Clean. Prod. **53**, 232–242 (2013)

C. Corbo, L. Lamastra, E. Capri, From environmental to sustainability programs: A review of sustainability initiatives in the Italian wine sector. Sustainability **6**(4), 2133–2159 (2014)

J. Costa, S. Catarino, J. Escalona, P. Comuzzo, Achieving a more sustainable wine supply chain: Environmental and socioeconomic issues of the industry, in *Improving Sustainable Viticulture and Winemaking Practices* (Academic Press, 2022), pp. 1–24

T. Da Silva Lopes, A. Lluch, G. Pereira, The changing and flexible nature of imitation and adulteration: The case of the global wine industry, 1850–1914. Bus. His. Rev. **94**(2), 347–371 (2020)

R. Dodds, S. Graci, S. Ko, L. Walker, What drives environmental sustainability in the New Zealand wine industry? An examination of driving factors and practices. Int. J. Wine Bus. Res. **25**(3), 164–184 (2013)

B. Figueroa, E. Rotarou, Challenges and opportunities for the sustainable development of the wine tourism sector in Chile. J. Wine Res. **29**(4), 243–264 (2018)

S. Flores, What is sustainability in the wine world? A cross-country analysis of wine sustainability frameworks. J. Clean. Prod. **172**, 2301–2312 (2018)

S. Flores, R. Medeiros, Wine tourism moving towards sustainable viticulture? Challenges, opportunities and tools to internalize sustainable principles in the wine sector, in M. Peris-Ortiz, M.C. Del Río Rama, C. Rueda-Armengot (eds.), *Wine and tourism: A strategic segment for sustainable economic development* (2016), pp. 229–245

S.L. Forbes, T. De Silva, A. Gilinsky, *Social sustainability in the global wine industry: Concepts and cases* (Palgrave MacMillan, 2020)

M.J. Gomes, A. Sousa, J. Novas, R. Jordão, Environmental sustainability in viticulture as a balanced scorecard perspective of the wine industry: Evidence for the Portuguese region of Alentejo. Sustainability **13**(18), 10144 (2021)

A. Goncharuk, Wine value chains: Challenges and prospects. J. App. Manag. Invest. **6**(1), 11–27 (2017)

J. Lichy, M. Kachour, P. Stokes, Questioning the business model of sustainable wine production: The case of French "Vallée du Rhône" wine growers. J. Clean. Prod. **417**, 137891 (2023)

G. Luzzani, L. Lamastra, F. Valentino, E. Capri, Development and implementation of a qualitative framework for the sustainable management of wine companies. Sci. Total. Environ. **759**, 143462 (2021)

A. Mariani, A. Vastola, Sustainable winegrowing: Current perspectives. Int. J. Wine Res. **7**, 37–48 (2015)

M. Millon, *Wine: A Global History* (Reaktion Books, 2013)

J.V. Montalvo-Falcón, E. Sánchez-García, B. Marco-Lajara, J. Martínez-Falcó, Sustainability research in the wine industry: A bibliometric approach. Agronomy **13**(3), 871 (2023)

A. Morrison, R. Rabellotti, Gradual catch up and enduring leadership in the global wine industry. Res. Policity **26**(2), 417–430 (2017)

A. Nave, A. Do Paço, P. Duarte, A systematic literature review on sustainability in the wine tourism industry: Insights and perspectives. Int. J. Wine Bus. Res. **33**(4), 457–480 (2021)

N. Ollat, J. Touzard, C. van Leeuwen, Climate change impacts and adaptations: New challenges for the wine industry. J. Wine Econ. **11**(1), 139–149 (2016)

M.A. Pratt, Comparison of sustainability programs in the wine industry. Bus. Ethics q. **14**(2), 243–262 (2012)

P. Signori, D.J. Flint, S. Golicic, Constrained innovation on sustainability in the global wine industry. J. Wine Res. **28**(2), 71–90 (2017)

J.P. Tiefenbacher, C. Townsend, The semiofoodscape of wine: The changing global landscape of wine culture and the language of making, selling, and drinking wine, in *Handbook of the Changing World Language Map*. ed. by S.D. Brunn, R. Kehrein (Springer, 2020), pp.4103–4145

J.M. Touzard, Y. Chiffoleau, C. Maffezzoli, What is local or global about wine? An attempt to objectivize a social construction. Sustainability **8**(5), 417 (2016)

E.C. Villanueva, G. Ferro, An update of the worlds of wine: The emerging countries' influence. Int. J. Econ. Bus. Res. **23**(1), 113–129 (2022)

Chapter 2
Theoretical Foundations

2.1 Sustainable Management Principles

The theoretical foundations of sustainable management and more particularly the consideration of the environmental, economic and social factor emerge from what is known as the triple bottom line concept from Elkington (1998), which constituted a revolution in the way of business performance evaluation by expanding the tradition financial assessment to a wider analysis by including social and environmental aspects.

The triple bottom line posits that true sustainability in business can only be achieved when equal importance is given to those three main dimensions: economic viability, environmental responsibility and social equity. In the context of the wine industry, this is translated by ensuring financial success (economic viability), minimizing environmental impacts in soil and water (environmental responsibility) and positively contributing to the communities in the areas where wine-related activities are developed (social equity) beyond the economic contributions linked to wine-producing itself.

Regarding the economic aspect, we must highlight that it goes beyond mere economic-financial ratios. It stands to long-term economic viability of the company, implying the well-being and stability of employees, suppliers and other close-related stakeholder relying on the industry. Specific manifestations of economic sustainability are (but not limited to) reduction of volatility by prioritizing stability, ensuring decent wages and promoting employee formation.

By its side, the environmental dimension urges businesses to surpass the care for the local nature beyond legal requirements by proactively making efforts to ensure the health of soil and water, the two main natural resources exploited by the sector. Water usage reduction, chemical product limitation and use of cover crops and organic materials help in that mission, favoring the preservation of local species and the viability of nature preservation. As Richter and Hanf (2021) pose, these actions can also lead to the sustainability of vineyards and the increase in the quality of

E. Sánchez-García et al., *Sustainable Management Through Knowledge and Innovation*, SpringerBriefs in Applied Sciences and Technology, https://doi.org/10.1007/978-3-031-64792-5_2

grapes, thus being an action that can also contribute to economic sustainability by the mentioned interrelations.

Finally, the social pillar emphasizes the importance of fair and beneficial practices that go beyond the economic activity itself. It is about the contribution that companies make to the society and groups they are related to. In this sense, the wine sector is not just embedded in certain areas, but they also depend on the sector. Supporting cultural events, culture initiatives or providing aid in areas such as education enhance this area of sustainability.

As we can see, contributing to each of the dimensions individually is easy to imagine. However, considering all three of them simultaneously is a great challenge. For example, organic farming may be a proper way for improving environment sustainability, but it also implies an increase in operating costs, thus reducing short-term profitability. The key element here are interrelations: some actions in favor of a certain dimension may be detrimental for other. In order to avoid such situation, adopting actions that create synergies for two or more dimensions (or, at least, that do not affect the other ones) is a must. Moving back to the example given: organic farming will increase operating costs, but if a proper business strategy via differenti-ation is developed, improving income more than costs is feasible. Another example would be investing in local education (social dimension), as it will enhance local workforce skills in the future, reverting the benefits derived from that action in the long term for wine production in the area (Ouvrard et al. 2020).

Therefore, the triple bottom line concept provides a robust framework for under-standing and implementing sustainable management in the wine industry, by giving equal importance to economic viability, environmental responsibility and social equity, so that wineries can enhance their sustainability while contributing to a more favorable future for the entire industry (Dressler and Paunović 2020). The successful integration of the triple bottom line approach requires a long-term perspective, inno-vative thinking and a commitment to continuous improvement and adaptation, as stated by Cataldo et al. (2021).

A proper example for achieving sustainability in the wine sector, in a broad sense, is circular economy. This practice consists of reintroducing waste and by-products in the supply chain, helping to the regeneration of natural systems by decreasing their exploitation. It can be applied to wine production, from vineyard management to bottling and distribution, considering that the wine sector is one of the most critical regarding pollutions as lots of by-products are generated (Musee et al. 2005). More specifically, we can highlight grape skins, stems and seeds (solid waste), this without including pH changes in possible surplus water emanating throughout the process. Nonetheless, this is indeed a challenge for cellars as considering the environmental impact of every aspect of their production process is needed, and so finding ways to minimize waste, reducing resource consumption and creating a more sustainable and efficient production cycle (De Steur et al. 2020). Among the different possibilities, D'Ammaro et al. (2021) mention some as creating compost from the aforementioned solid by-products (thus reducing the use of synthetic fertilizers) and grapeseed oil (which is a high-value good).

Focusing now on water exploitation, this is a precious resource in viticulture, as 5000 to 6000 L are needed to produce 1 L of wine (Bujdosó and Waltner 2017). Thus, its efficient use and reuse are essential for sustainable wine production, so that implementing systems for collecting and reusing wastewater in irrigation, adopting water-efficient technologies like drip irrigation and designing wineries with the focus on optimizing water use are all practices that align with the circular economy model, being this particularly important in regions prone to drought and water scarcity (Cabrera-Flores et al. 2020).

Another important action where the wine industry has to improve is energy consumption, as transitioning to renewable energy sources and implementing energy-efficient technologies can reduce the carbon footprint of wine production. The main issue is that the wine industry relies on activities from different nature (agricultural, industrial and transportation), so energy sources are varied and, on many occasions, difficult to be replaced by more efficient counterparts. However, and despite the initial investment required for implementing new technologies and practices in this sense, it can be offset by long-term benefits as revenue from by-products and higher operational optimization, something that is mentioned in Muñoz et al. (2020).

Indeed, circular economy offers a transformative approach to sustainable management in the wine industry, providing a framework for reducing environmental impact, optimizing resource use and creating a more sustainable and efficient production cycle, so that by embracing its principles, wineries can contribute to a more sustainable future for the industry (and the environment) while creating a competitive advantage in a market that is increasingly focused on sustainability (Ferrer et al. 2022). The design and implementation of circular economy practices in the wine industry requires innovation, investment and a commitment to rethinking traditional practices, what undoubtedly are actions that require a considerable amount of time and diverse capabilities, but the potential benefits for the environment, the economy and society make it a crucial component of sustainable management (Andrade-Suárez and Caamaño-Franco 2020).

However, we can state that the most important element for sustainability, not only in the wine industry but in any economic activity, is corporate social responsibility (CSR), which expands the goals of a business further than mere economic profit, considering the impact it has in other terms, particularly the environment and society. In this sense, and as it can be deduced, most of the actions exposed in the previous are, in fact, the materialization of CSR practices. Considering all that has been previously stated, is obvious that CSR requires a commitment to long-term goals, often involves additional costs and investments, but it helps to contribute to the overall health and well-being of the environment and society, which is essential for the long-term viability of the wine industry. Thus, by integrating corporate social responsibility into their core business strategy, wineries can enhance their sustainability, build trust and loyalty among stakeholders and contribute to the overall health and well-being of the environment and society (Svanidze and Costa-Font 2023).

2.2 Innovation in the Wine Industry

Disruptive innovation refers to a process by which a product or service takes root initially in simple applications at the bottom of a market and then relentlessly moves upmarket, eventually displacing established competitors, being particularly pertinent to the wine industry due to its traditional practices and market dynamics challenged by innovative approaches and technologies (Frigon et al. 2020). In the context of the wine industry, disruptive innovation can be observed in various forms, e.g., the advent of new wine production technologies that have significantly altered traditional viticulture and winemaking processes, precision viticulture (which uses GPS and satellite data to optimize vineyard management) or advanced fermentation techniques that enhance flavor profiles, which represent disruptive changes to the conventional methods of wine production, which improve efficiency and product quality and challenge the traditional practices that have defined the industry for centuries (Angelova and Pastarmadzhieva 2020a).

Another form of disruptive innovation in the wine industry is seen in the business models and distribution channels, where the rise of direct-to-consumer sales models (strongly facilitated by online platforms and social media) has displaced the traditional wine distribution channels in certain contexts. The main benefit it brings, if compared to older practices, is a more direct relation with customers and a personalized marketing strategy, both of which enable smaller wineries to compete effectively against larger, more established companies (Dressler and Paunovic 2021).

Moreover, the emergence of new wine styles, among which we can mention organic, biodynamic or natural ones, challenges the traditional wine offer and is generating a growing segment of consumers who are looking for unique and sustainable wine options, which often require different production methods and marketing approaches, that can be considered as a disruption of the conventional practices, thus leading to innovation and differentiation (Rabadán 2021). Nonetheless, the application of such innovations requires responsive wineries to be responsive to changes in the competitive context, so they can challenge the *status quo* regarding the ways of doing business (Doloreux et al. 2020). However, these innovations involve risks linked to the significant investments required and the radical changes they imply for the sector (Kumar et al. 2022). Being the wine industry strongly underpinned in history and tradition, radical innovations may not be easily accepted by many consumers, so that wineries need to carefully consider how to integrate innovation with the heritage of the industry (Morata et al. 2023), so that new customer segments are created without alienating the current customer base.

Another possibility is the so-called open innovation, a term coined by Chesbrough (2003) to describe an approach where businesses rely both on their internal ideas and resources and on external sources of technology, talent and knowledge for developing innovation. In this sense, "open" is indeed "openness" for external potential proposals that are developed by innovation-driven organizations. In the context of the wine industry, open innovation allows wineries to benefit of a more varied melting pot of ideas by collaborating with research institutions and universities, startups and other private technological companies. By doing so, wineries have access to the latest scientific developments, innovative technologies and novel business practices that might not have been able to develop internally.

This innovation can also be developed hand in hand with clients and suppliers. On the one hand, collaboration with customers can be achieved through platforms that gather customer feedback for elements like new wine varieties or packaging designs, and also by engaging them to the product by wine tastings, with wineries gathering valuable insights of their preferences, leading to the development of products closely aligned with market demands and, at the same time, creating a deeper connection between wineries and customers (Richter and Hanf 2021). On the other hand, innovation by collaboration with suppliers can be achieved through the materialization of the prior proposals, as they become key partners in creating the best value-added offer. Anyways, the implementation of open innovation requires a shift from a traditional, closed approach to innovation to one that is more collaborative, aimed at building networks and partnerships with a variety of external entities, some of them maybe not closely related to the sector, something that can be challenging given the different "realms" embedded in the wine industry culture. To develop such innovation, ensuring and making explicit that collaborations and partnerships are mutually beneficial is crucial, something that may involve the adaptation of different organizational structures and processes to make that collaboration possible (Roudil et al. 2020). A good materialization may take place by the known as value innovation, a concept developed by Chan Kim and Mauborgne (2005) which focuses on the idea of breaking away from intense competition by creating new market spaces, defined as blue oceans, where the competition is irrelevant as, in fact, it does not exist, so rather than beating the competition, the company would make the competition irrelevant by creating value for both the company and its customers. Examples of open innovation in the wine sector include direct-to-consumer business models, subscription services and online education programs, all of them brought from other sectors and creating new business opportunities (Angelova and Pastarmadzhieva 2020b).

Another possibility that streams in this topic is incremental innovation. Unlike the previous ones, which are characterized by a deep break with the current state, incremental innovation is closer to the present reality, meaning that small, progressive changes are done in product, processes and/or business models. Incremental innovations have small-scale improvements as an outcome and are crucial in industries with high sensitiveness, where small changes can lead to significant alterations in its structure and functioning (Faria et al. 2020), being the wine industry such a case. Thus, we can say that incremental innovation is seen as an action toward adjustment that helps to improve the quality and distinctiveness of wine-related products and services over time (Martínez-Falcó et al. 2023a). Despite its more accessible and less hazardous development, incremental innovation also has a considerable drawback: it can be less visible and harder to communicate to consumers compared to more radical innovations, as changes are subtler.

Regardless whether radical or incremental innovations, an interesting option given the current demands that customers exert on the industry is the so-called green innovations. We refer to green innovation as the integration of sustainability into the innovation process, focusing on creating products, processes and business models that are economically viable, environmentally responsible and socially beneficial (i.e., that they comply with the three dimensions of sustainability). This sort of innovation, as Bell and Giuliani (2007) indicate, requires a long-term vision and considering the full lifecycle impact of their actions, from vineyard management to wine production, packaging, distribution and consumption. Sustainable innovation in the wine sector can be also adopted by the introduction of non-specific sectorial changes. For example, projects aiming at implementing energy-efficient and low-impact processes, as self-consumption energy sources or optimized procedures in the distribution phase (which implies transportation). Given the use of more business-to-customer (B2C) models, lots of wineries have implemented their own distribution channels to reach the final consumer. Apart from transportation, packaging is another way to approach sustainable innovation. Creating lighter recipients, reintroducing by-products (for example, in labels and cork coatings) among others, so the global impact of the product is minimized. Besides reducing the environmental impact of the wine production and commercialization, these initiatives also contribute to cost savings and the winery's resilience to the fluctuation in the price of inputs.

Summing up, different ways to develop innovation are within the reach of wineries. What is clear is that, by the increasing demands of customers regarding more sustainable, differentiated and quality products, the industry has to properly respond to those requirements, applying the innovations that best fit the needs and the reality of the locations where they develop business. The rapid technological transformations that are happening have come together with a need for the adaptation of economic sectors. Being the wine industry strongly rooted in its history, a considerable effort has to be made to change its vision, guaranteeing its long-term health while integrating history with innovation, something that requires a different perspective than the one applied up to date. Thus, as a conclusion, and without any aim of excluding other possible types of innovation, Fig. 2.1 summarizes the typologies of innovation that emerge from the combination of the different types explained, where the easier to materialize will be more likely situated in the coordinate origin, being harder to develop the further from it the innovation is located.

2.3 Integrating Knowledge and Innovation: How Knowledge Management Supports Innovation Management

Knowledge management refers to the actions toward gathering, refining and distributing information (Benjamins et al. 1998) that is relevant for the strategy of a business. Knowledge management also includes its sharing, being it equally important. Nowadays, activities such as conferences, research publication, collaborative research projects and informal networks serve as places to share knowledge. One of the main challenges that companies in the wine industry face in this sense is the

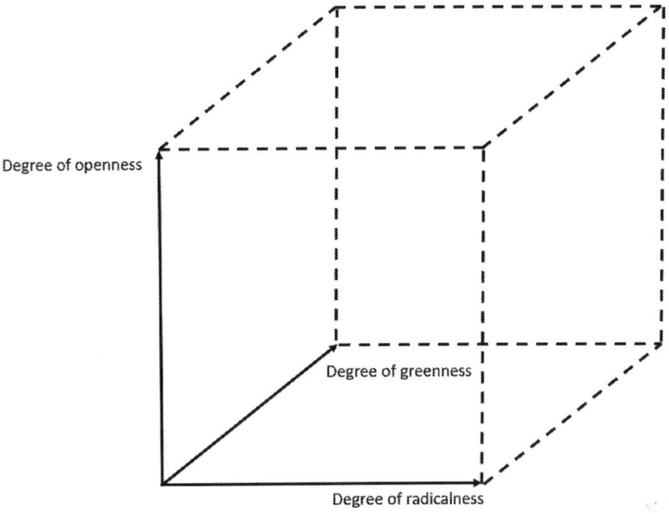

Fig. 2.1 Different combinations of innovations. *Source* own elaboration

barriers for knowledge exchange, as Doloreux et al. (2015) point out. These barriers hinder the transfer of information between companies, and even inside firms. Alonso Ugaglia and Peres (2017) state that a commitment to collaboration and openness together the development of knowledge-sharing mechanisms is needed if there is a real pretension to surpass these barriers. Another challenge is ensuring that the knowledge created and shared is relevant for the company and applicable to the industry (Lenzi 2013). This requires a proper identification of strengths and weaknesses within the company, as well as the threats and opportunities that come from the industry. It also needs for engaging a wide range of participants, as researchers, practitioners and consumers are important actors for knowledge creation and transmission. In this vein, the integration of creation and sharing is costly in terms of effort, but the outcomes derived from them are an essential component for sustainable management, also in the wine industry (Galbreath 2015).

Related to knowledge management, organizational learning is constituted as one of its main pillars. It is an activity that demonstrates the need for companies to acquire external information continuously and apply it in order to gain sustainable advantages, being useful to face environmental changes, evolving market demands and technological developments (Signori et al. 2017). Thus, we can say that organizational learning is the ability of a firm to learn from its successes and failures, what needs of a culture that sees this highs and lows as an opportunity for learning and growth (Choi and Gu 2020). Indeed, the ability to learn from external sources involves continuous updating in the latest technologies, practices and novel research, something that is inseparably accompanied by uncertainty and high risks. This uncertainty, in the case of the wine industry, is reflected by considering that very different stakeholders take part in this activity, as research institutions, industry associations and wide networks of wine producers and experts are needed. However, this "complexity" also brings new insights in sustainability practices, consumer trends and innovative technologies (Lekics 2021).

When talking about organizational learning, tacit knowledge and its integration within the company cannot be avoided. Coined by Polanyi (1958), it refers to the knowledge that is based on personal experience and learning that is difficult to transfer by means of formalization (such as writing or verbalizing), so it is embedded and indivisible from its creator. Things as the particularities of grape cultivation and the sensory evaluation of wine, which are learnt through years of experience, are examples of that.

Despite knowledge management is an unquestionable boost for tacit knowledge, its translation into sustainable innovation encompasses certain difficulties. First of all, a practical and deep understanding of the crafting process is needed, as it cannot be easily codified and documented, as it is affected by many variables that are uncontrolled by vine-growers and winemakers. Understanding microclimates, soil types, grapevine behavior and their interactions requires a level of intuition and experience that transcends formal knowledge (Dries et al. 2013). One way to overcome this barrier is through mentorship and apprenticeship programs, so that experienced winemakers and viticulturists interact with novel personnel and transmit part of their knowledge while enhancing self-creation for the latter. That way, new generations will create and acquire knowledge upon the wisdom and experience of their predecessors.

Another way for integrating tacit knowledge is through collaborative experimentation and problem-solving. This approach reunites individuals with diverse skills and experiences and promotes cost sharing along the process. This is particularly critical in the case of sustainable management, as the developments and implementation of new technologies and practices is a time- and resource-consuming task, so by collaborating, those costs are partly mitigated.

Independently from the above, the integration of knowledge and innovations requires of a culture that values and encourages knowledge sharing, i.e., an environment where expressing their insights and experiences is rewarded, where the willingness to learn is a constant and where mistakes are not seen as something to be punished for. Inasmuch as tacit knowledge is inherently personal and context-specific, its transfer and replicability are partially impeded, so the creation of opportunities for direct interaction (for example, by group dynamics) is a proper way for facilitating knowledge spread.

In this regard, it is also important to know how knowledge management supports sustainable innovation in the wine industry by means of knowledge codification and transfer. Knowledge codification is defined as the process of converting tacit knowledge (at least, as much as it is plausible due to its nature) into explicit knowledge that can be consequently documented, recorded and shared, being especially useful for its transfer among different stakeholders over time, and involves documenting the insights, techniques and practices that have been developed over time, such as detailed records of viticulture practices, winemaking processes and the outcomes of various experiments. For example, a winery that keeps a record of different fermentation processes and their effects on wine flavor, creating a valuable resource that can be used for training, decision-making and innovation would be an example, as reflected in Martínez-Falcó et al. (2023a). Once the knowledge has been codified, it must be disseminated to the relevant stakeholders within and outside the organization, for which training programs, wine conferences and similar group events can serve to that purpose, as they enable wineries to detect and implement the best practices and innovations (Giuliani 2008). Once this sharing has been accomplished, that virtuous cycle of acquisition, processing, assimilation and sharing begins again. The prior is not just a theoretical reflection. The aim for wineries of recording their practices and outcomes has let new cellars create their own techniques based on previous knowledge, thus focusing their efforts on experimenting with new ideas, as part of the work had been done by those who preceded them. So, we can say that ensuring the creating of new knowledge, ensuring its transfer and enhancing its capture are a must for wineries that aim to push their boundaries, for what clear documentation and effective communication are effective. Given the nature of sustainable innovation, this probably becomes the best way to achieve it in the industry (Galbreath et al. 2016).

We should not end this chapter without talking about the knowledge networks that make knowledge transfer possible, which are also indispensable for developing a proper management of sustainability. These networks are composed by many different stakeholders: vineyard managers, winemakers, suppliers, distributors, customers and competitors, as well as other actors that are not directly linked to the sector but can influence in it, such as researchers and policymakers. This variety of stakeholders is one of the reasons why knowledge dissemination is a complex venture: their differences in social, technical and economic terms require an adaptation for each of them. At the same time, is that variety what enhances knowledge creation, so it is a complex but necessary reality to face.

Being some of the previous collaborations present in knowledge networks intuitive, we have to have a look at collaboration with competitors. Often referred to as co-opetition, rather than being an impediment for the growth and development of wineries, working together with rivals in joint projects can reduce the intrinsic risk of those ventures and may reduce hostility among collaborators (if the agreement is properly managed). Examples of such collaborations are the so-called Designations of Origin, where different wineries have to work together for the recognition of the wines from their location as being differentiated than products from other sites. However, we also have to keep in mind that collaboration (and particularly with competitors) also brings certain complexities, as divergence of goals, expectations and working styles, for what effective coordination, communication and protocols are needed to control those deviations from collective benefit.

References

A. Alonso Ugaglia, S. Peres, Knowledge dynamics and climate change issues in the wine industry: A literature review. J Innov Econ Manage **1**(16), 21–36 (2017)

M. Andrade-Suárez, Caamaño, I. Franco, The relationship between industrial heritage, wine tourism, and sustainability: A case of local community perspective. Sustainability **12**(18), 7453 (2020)

M. Angelova, D. Pastarmadzhieva, Conceptual model for investigating the innovation activity of enterprises. Sci. Bus. Soci. **5**(1), 17–19 (2020a)

M. Angelova, D. Pastarmadzhieva, Development of bio-based economy: Entrepreneurial endeavors and innovation across Bulgarian wine industry. J. Int. Stud. **13**(2), 149–162 (2020b)

M. Bell, E. Giuliani, Catching up in the global wine industry: Innovation systems, cluster knowledge networks and firm-level capabilities in Italy and Chile. Int. J. Technol. Glob. **3**(2–3), 197–223 (2007)

R. Benjamins, D. Fensel, A. Gómez-Pérez, Knowledge management through ontologies, in *2nd International Conference on Practical Aspects of Knowledge Management* (1998)

B. Bujdosó, I. Waltner, Water footprint assessment of a winery and its vineyard. Hungarina Agric. Res. **26**(1), 10–13 (2017)

M. Cabrera-Flores, J. López-Leyva, M. Peris-Ortiz, A. Orozco-Moreno, J. Francisco-Sánchez, O. Meza-Arballo, A framework of penta-helix model to improve the sustainable competitiveness of the wine industry in Baja California based on innovative natural resource management. E3S Web Conf. **167**, 06005 (2020)

E. Cataldo, M. Fucile, G.B. Mattii, A review: Soil management, sustainable strategies and approaches to improve the quality of modern viticulture. Agronomy **11**(11), 2359 (2021)

W. Chan Kim, R. Mauborgne, Value innovation: A leap into the blue ocean. J. Bus. Strateg. **26**(4), 22–28 (2005)

H. Chesbrouugh, *Open innovation: The new imperative for creating and profiting from technology* (Harvard University Press, 2003)

H. Choi, C. Gu, Geospatial response for innovation in the wine industry: Knowledge creation through institutional mobility in China. Agronomy **10**(4), 495 (2020)

D. D'Ammaro, E. Capri, F. Valentino, S. Grillo, E. Fiorini, L. Lamastra, A multi-criteria approach to evaluate the sustainability performance of wines: The Italian red wine case study. Sci. Total. Environ. **799**, 149446 (2021)

H. De Steur, H. Temmerman, X. Gellynck, M. Canavari, Drivers, adoption, and evaluation of sustainability practices in Italian wine SMEs. Bus. Strateg. Environ. **29**(2), 744–762 (2020)

D. Doloreux, R. Shearmur, R. Guillaume, Collaboration, transferable and non-transferable knowledge, and innovation: A study of a cool climate wine industry (Canada). Growth Chang. **46**(1), 16–37 (2015)

D. Doloreux, R. Shearmur, I. Porto-Gomez, J.M. Zabala-Iturriagagoitia, DUI and STI innovation modes in the Canadian wine industry: The geography of interaction modes. Growth Chang. **51**(3), 890–909 (2020)

M. Dressler, I. Paunović, Converging and diverging business model innovation in regional intersectoral cooperation—Exploring wine industry 4.0. Eur. J. Innov. Manag. **24**(5):1625–1652

M. Dressler, I. Paunović, Towards a conceptual framework for sustainable business models in the food and beverage industry: The case of German wineries. Br. Food J. **122**(5), 1421–1435 (2020)

L. Dries, S. Pascucci, Á. Török, J. Toth, Open innovation: A case-study of the Hungarian wine sector. EuroChoices **12**(1), 53–59 (2013)

J. Elkington, Accounting for the triple bottom line. Meas. Bus. Excell. **2**(3), 18–22 (1998)

S. Faria, L. Lourenço-Gomes, S. Gouveia, J.F. Rebelo, Economic performance of the Portuguese wine industry: A microeconometric analysis. J. Wine Res. **31**(4), 283–300 (2020)

J. Ferrer, M. García-Cortijo, V. Pinilla, J.S. Castillo-Valero, The business model and sustainability in the Spanish wine sector. J. Clean. Prod. **330**, 129810 (2022)

A. Frigon, D. Doloreux, R. Shearmur, Drivers of eco-innovation and conventional innovation in the Canadian wine industry. J. Clean. Prod. **275**, 124115 (2020)

J. Galbreath, To cooperate or compete? Looking at the climate change issue in the wine industry. Int. J. Wine Bus. Res. **27**(3), 220–238 (2015)

J. Galbreath, D. Charles, E. Oczkowski, The drivers of climate change innovations: Evidence from the Australian wine industry. J. Bus. Ethics **135**, 217–231 (2016)

E. Giuliani, What drives innovative output in emerging clusters? Evidence from the wine industry. SPRU Electron. Working Paper Ser. **169**, 1–39 (2008)

M. Kumar, M. Pullman, T. Bouzdine-Chameeva, V. Sanchez Rodrigues, The role of the hub-firm in developing innovation capabilities: Considering the French wine industry cluster from a resource orchestration lens. Int. J. Oper. Prod. Manag. **42**(4), 526–551 (2022)

V. Lekics, Sustainable innovation in wine industry. A systematic review. Regional Bus. Stud. **13**(1), 55–73 (2021)

C. Lenzi, Smart upgrading innovation strategies in a traditional industry: Evidence from the wine production in the province of Arezzo. Reg. Sci. Policy Pract. **5**(4), 435–452 (2013)

J. Martínez-Falcó, B. Marco-Lajara, P. Zaragoza-Sáez, L. Millán-Tudela, Wine tourism as a catalyst for green innovation: Evidence from the Spanish wine industry. Br. Food J. **126**(5), 1904–1922 (2023a)

J. Martínez-Falcó, E. Sánchez-García, B. Marco-Lajara, L.A. Millán-Tudela, Do organizational commitment and consumer satisfaction mediate the relationship corporate social responsibility-sustainable performance? Assessing happiness management in Spanish wineries. Manag. Decis. **62**(2), 643–664 (2023b)

A. Morata, T. Arroyo, M. Bañuelos, P. Blanco, A. Briones, J. Cantoral, V. Capozzi, Wine yeast selection in the Iberian Peninsula: Saccharomyces and son Saccharomyces as drivers of innovation in Spanish and Portuguese wine industries. Crit. Rev. Food Sci. Nutr. **63**(31), 10899–10927 (2023)

R. Muñoz, M. Fernández, Y. Salinero, Sustainability, corporate social responsibility, and performance in the Spanish wine sector. Sustainability **13**(1), 7 (2020)

N. Musee, L. Lorenzen, C. Aldrich, Decision support for waste minimization in the wine-making process. Environ. Prog. **25**(1), 56–63 (2005)

S. Ouvrard, S. Jasimuddin, A. Spiga, Does sustainability push to reshape business models? Evidence from the European wine industry. Sustainability **12**(6), 2561 (2020)

M. Polanyi, *Personal Knowledge: Towards a Post-Critical Philosophy.* Routledge (1958)

A. Rabadán, Consumer attitudes towards technological innovation in a traditional food product: The case of wine. Foods **10**(6), 1363 (2021)

B. Richter, J. Hanf, Cooperatives in the wine industry: Sustainable management practices and digitalization. Sustainability **13**(10), 5543 (2021)

L. Roudil, P. Russo, C. Berbegal, W. Albertin, G. Spano, V. Capozzi, Non-Saccharomyces commercial starter cultures: Scientific trends, recent patents and innovation in the wine sector. Recent Pat. Food Nutr. Agric. **11**(1), 27–39 (2020)

A. Svanidze, M. Costa-Font, Revealing the challenges facing Georgia's wine industry from a natural winemaker perspective using Q-methodology. Int. J. Wine Bus. Res. **35**(1), 89–120 (2023)

P. Signori, D. Flint, S. Golicic, Constrained innovation on sustainability in the global wine industry. J. Wine Res. **28**(2), 71–90 (2017)

Chapter 3
Developing Sustainable Business Strategies

3.1 Strategic Assessment of Sustainability

In the prior chapter, we ended up by showing the dynamics of the integration between knowledge and sustainability. However, knowledge is just one of the many tools at the disposal of the wine sector to achieve sustainability. The different ways to act are taken up in the strategy of wineries. More specifically, strategic planning for sustainability is a critical process for businesses, even more in the wine industry given the impact it has on the environment and the considerable dependence existing on it. The most popular way to consider the need for sustainability in the strategy of a business is by the already presented triple bottom line proposed by Elkington (1998), as it expands the traditional financial reports to include environmental and social performance. This approach allows to develop an integral strategy that considers all the aspects where a business affects as a consequence of the activity it develops. As it is also part of the analysis, formulation and implementation of a firm's strategy, it is present in every phase that is required.

However, the consideration of the different dimensions is not something that is limited to the aim to contribute to sustainable development. Many companies do what is known as "greenwashing", which is the deliberate corporate action with the presence of misleading elements, focused on the deception of stakeholders (de Freitas Netto et al. 2020), all of this related to sustainability. Thus, there is a need to create and report a clear and measurable set of goals that can be objectively assessed not only by the company, but also for the different stakeholders. In this regard, wineries need to develop specific strategies and action plans accompanied by congruent and objective measurement tools that try to assess the achievement of goals that have not been underestimated. These strategies, actions and measurement tools have to be targeted toward the improvement of sustainability by means of improving knowledge and skills, ensuring a proper allocation of resources, and measuring compliance by part of the organization members, all of these by establishing realistic but ambitious milestones. From all these activities, monitoring and reporting are the most critical (or, at least, one of the most) in the context of the triple bottom line. Reporting should be both internal (informing to management and employees) and external (thus communicating customers, investors and other stakeholders the actions and results in the context of sustainability), for what proposals as the Global Reporting Initiative (2024) or the AA1000 Accountability Stakeholder Engagement Standard from the International Organization for Standardization (2015) can be an interesting tool. This will provide a comprehensive view of the winery's impact and performance to all interested parts, allowing wineries to make more informed decisions that balance short-term needs with long-term sustainability goals, thus helping wineries identify new opportunities for innovation and growth.

One of the most applied tools for the control and subsequent report of that information is the known as balanced scorecard developed by Kaplan and Norton (1991). This technique consists of a series of four dimensions (financial, customer, internal business and innovation and learning), for which the objectives must be explicated. From the first version of the tool, different add-ons have been developed. Probably the most interesting one is the inclusion of indicators and milestones together with time targets. By doing so, a company is able not only to show what they pretend to achieve, but also the way those goals are about to be measured, as well as their objectification. Being the balanced scorecard a general tool that is intended to translate the theoretical strategy for a company into operational activities and objectives (Bochenek 2019), we can apply it to the main purpose of sustainability in the wine industry, for what each of the main dimensions is going to be explained considering that framework.

From the financial perspective, the balanced scorecard can be used by wineries to express and assess the way sustainability initiatives can contribute to long-term financial performance, which entails analyzing the cost savings from resource efficiency improvements, the revenue potential of sustainable wine products or the financial risks associated with environmental and social factors. In this stream, we can say that financial goals related with sustainability must be accompanied by a series of economic and financial ratios that express their efficiency and compare them with their less-sustainable counterparts as means of transparency and proper strategic assessment.

The customer perspective focuses on understanding the expectations of customers in order to meet them. In the case of sustainability, it can be used to enhance the actions toward the communication of the sustainability actions done by the firm, improving sustainable wine products by considering consumer preferences, and engaging customers in sustainability initiatives. In a broader sense, by means of the customer perspective, companies can control if the actions that have developed are having a real impact in the satisfaction of their clients.

Regarding internal processes perspective, it is directed toward assessing the functioning of the routines and activities that, from an operational point of view, are developed within companies. It not only measures how processes are done, but also the outcomes obtained with them. For the case of sustainability in the wine sector, it can be applied to evaluate the performance of operations, the degree of sustainability of the different activities or the reduction in the impact they can have compared to prior operations.

Finally, we find the learning and growth perspective, which is directly linked with the development of a company's personnel, systems and procedures. It is directed toward facilitating the measurement, assessment and consequent feedback about the impact of training initiatives and the performance of current systems and procedures. In this sense, it can be used to check the impact of possible sustainability formations for employees, the sensitiveness of systems related to sustainability and the precision of the procedures in this area.

However, measuring the sustainable success of a company with the balanced scorecard requires that firms have a clear vision of it, so that goals and consequent indicators can be fixed. It also requires of a proper infrastructure that allows the constant monitoring of the desired variables, and ensuring that these measures are relevant and representative of the real situation and needs regarding sustainability. Reflecting on needs, one of the most common mistakes is to establish milestones that are unambitious or extremely far in time, so actions toward sustainability are not strongly deployed or delayed, respectively. Thus, we can say that the challenge is to ensure that sustainability objectives are aligned with the real needs of companies in the sector, that the measurements and the mechanisms to apply them are properly designed and that the levels to reach and the time available are congruent with the true aims of the firm. Nonetheless, we can state that the balanced scorecard is an easy-to-apply but profound tool that is applicable to any kind of firm within the wine industry and is a proper propeller of sustainability if it is well designed and materialized.

3.2 Stakeholder Engagement: The Importance of Involving Stakeholders in Sustainability Initiatives

Involving stakeholders in sustainability initiatives is a critical aspect in present firm strategies, and even more importantly in industries like wine production, where the link with the environment is profoundly rooted in nature, having a strong interdependence with inhabitants of the locations where the activity is established, and being in many times a powerful economic driver of such areas. In the case of the wine industry, stakeholders are a diverse group of individuals and organizations, including (but not limited to) vineyard workers, local communities, suppliers, customers, regulatory bodies, environmental groups and investors. Consequently, sustainability challenges and opportunities are not limited to economic aspects, as these stakeholders also impose social and environmental claims to the industry. Having each of these stakeholder groups unique insights, concerns and expectations, making their involvement in achieving sustainability is crucial, as the potential divergence of demands requires of a profound knowledge of each of these groups. By engaging with these stakeholders, wine-related firms can have specific insights from each of them and a proper image of their needs, thus leading to more effective and socially responsible practices.

Customers are currently concerned about the impact of the products they purchase, so making them part of sustainability initiatives can serve as a way for enhancing brand reputation and loyalty (Akbari et al. 2020), while also providing valuable information about market trends. The same can be stated for investors, as receiving their claims in regard to sustainability can enhance decision-making toward sustainability, and thus getting closer to such demands. Proof of that is the existence of stock indexes directed toward sustainability, such as the FTSE4Good Index Series (London Stock Exchange Group 2024) or the Dow Jones Sustainability Index (S&P Dow Jones Indices 2024), among others.

Therefore, implementing effective actions aimed at stakeholders' engagement requires of a proper identification of who key stakeholders are, as well as understanding their concerns and expectations to design proper mechanisms for collaboration and information transparency with them. However, balancing the diverse (and sometimes incompatible) interests of different stakeholder groups can be a great challenge, requiring of effective communication, negotiation and, more importantly, an innovative approach to those conflicts of interest that emerge, so that most of the claims can be satisfied without leaving any of the key stakeholders unserved.

In connection with the above, involving stakeholders in sustainability initiatives in the context of the wine industry requires awareness of the fact that stakeholder engagement goes beyond simple consultation. A proper engagement also includes a constantly collaborative environment where stakeholders' insights and experiences actively contribute to the formulation and implementation of sustainability strategies. The sequence of actions is similar to the so-called high-level strategies: it begins with the identification and mapping of stakeholders (analysis of involved actors), being followed by the design (formulation) and the subsequent deployment of communication channels with them (implementation), acquiring the form of meetings, workshops or surveys, among other possibilities. These channels will allow wine-related firms to gather the required information (analysis) of stakeholders' sustainability demands, which allows for the formulation of sustainable actions that, in the end, become applied. Figure 3.1 shows the loop of actions made by the sequence of activities aimed at stakeholder engagement.

Thus, we can say that effective stakeholder engagement in sustainability initiatives leads to more informed and robust sustainability strategies, as they incorporate the insights and experiences of the stakeholders that, if not properly considered, can overturn those strategies. Moreover, it guarantees their acceptance, as stakeholders are actively involved in the development and implementation of sustainability strategies. Finally, effective stakeholder engagement can lead to stronger partnerships through the establishment of long-term relationships that are based on mutual respect and shared goals.

While the benefits of engaging stakeholders are clear, the process to do so can be complex and tough when brought to practice. Understanding the real needs of each group and put them all together (considering that they may be incompatible)

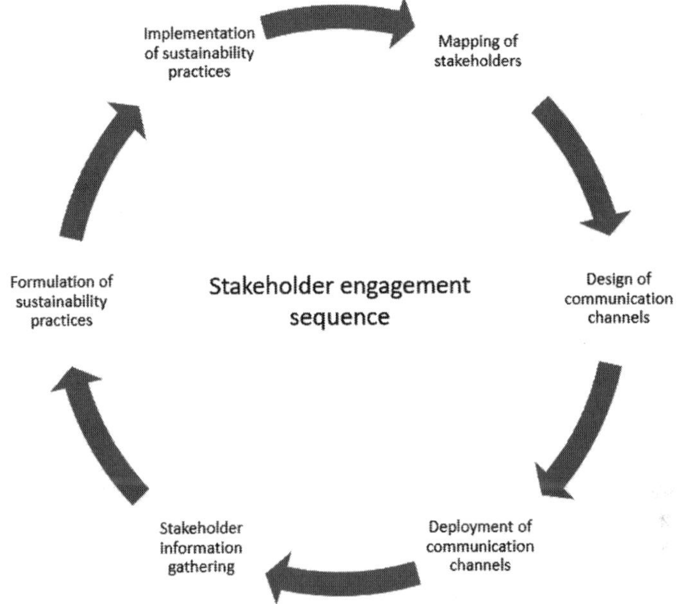

Fig. 3.1 Stakeholder engagement sequence. *Source* own elaboration

requires of political abilities, where persuading each of the groups to give in on some of their claims to merge the unavoidable needs together in the sustainability strategies of wine-related firms is mandatory, as the work of Velter et al. (2020) hints. This becomes nothing but even more complex considering the wide range of stakeholders and, consequently, of claims: local community members, employees, customers and suppliers, regulators, environmental groups and investors, not to mention other potential groups. Effective stakeholder engagement also requires of significant amounts of time and resources, that have to be retrieved from other areas of the firm, so controlling for resource-allocation optimization is crucial. Despite the challenge it means, its importance in achieving sustainable outcomes and building strong, resilient businesses in the wine industry cannot be overstated, so stakeholder engagement in the context of sustainability cannot be ignored if a resilient, long-term subsistence of the wine industry is truly aimed.

References

M. Akbari, M. Mehrali, N. SeyyedAmiri, N. Rezaei, A. Pourjam, Corporate social responsibility, customer loyalty and brand positioning. Soc. Res. J. **16**(5), 671–689 (2020)

M. Bochenek, Balanced scorecard in strategic management process. Mod. Manag. Rev. **24**(26), 7–16 (2019)

S.V. De Freitas Netto, M.F. Falcão Sobral, A.R.B. Ribeiro, G.R. Da Luz Soares, Concepts and forms of greenwashing: A systematic review. Environ. Sci. Eur. **32**(19), 1–12 (2020)
J. Elkington, Accounting for the triple bottom line. Meas. Bus. Excell.excell. **2**(3), 18–22 (1998)
Global Reporting Initiative, *About GRI* (2024, May 1). https://www.globalreporting.org/about-gri/
International Organization for Standardization, *AA1000 Stakeholder Engagement Standard* (2015)
S. Kaplan, D. Norton, The balanced scorecard: Measures that drive performance. Harvard Bus. Rev. January-February, 71–79 (1991)
London Stock Exchange Group, *FTSE4Good Index Series* (2024, May 3). https://www.lseg.com/en/ftse-russell/indices/ftse4good
S&P Dow Jones Indices, *Dow Jones Sustainability Index* (2024, May 3). https://www.spglobal.com/spdji/en/indices/esg/dow-jones-sustainability-world-index/#overview
M.G.E. Velter, V. Bitzer, N.M.P. Bocken, R. Kemp, Sustainable business model innovation: The role of boundary work for multi-stakeholder alignment. J. Clean. Prod. **247**, 119497 (2020)

Chapter 4
The Learning Organization as a Propeller of Innovation and Sustainability

4.1 The Concept of Learning Organization

In the prior chapters, we have delved into the concepts of innovation and sustainability in the wine industry, explaining the need for firms in this sector to be sustainable and the role that innovation plays on its achievement. We have also seen how the firm interacts with the environment in order to achieve that goal, and even some brief concepts regarding the configuration of businesses to be aligned toward innovation and sustainability have been presented. However, a profound reflection on the real set-up a firm needs both in terms of innovation and sustainability is needed, as staying on the surface would limit the application of all of the previously mentioned. However, we must warn that the aim of this chapter is not to specify in a deterministic way how firms in the wine sector should configure themselves. The following lines have to be interpreted as indications for those who want to initiate in the transformation of their wine-related businesses to improve their chances of improving sustainability via innovation. Consequently, we will provide certain indications, based on what has been exposed in previous chapters, so that practitioners and other professionals related to the industry have a starting point to introduce themselves in this adventure.

First, we have to expose the term "learning organization". According to de Geus (2002), a learning organization is the one which is constantly changing by means of acquiring and developing new knowledge, for what openness to the environment, avoiding punishment to mistakes and an alignment of interest with all its members must be present.

As de Geus points out, the first feature of a learning organization is its inherent dynamicity, i.e., that it is constantly changing. He identifies such firms as "river companies" compared to "pool (static) companies". Avoiding literary prose, the fact that learning companies are conceived as of constantly changing implies a certain inner configuration that allows such behavior. The most obvious is the need for a horizontal structure, that reduces the formal distance between the different levels within the firm, what enhances communication among the members of the organization (Khatri 2009) and, thus, a greater generation and spread of knowledge. Nonetheless, this structure cannot be implemented alone, as it needs accompaniment of other related characteristics. For example, information within the company (with the exception of sensitive data) has to be available for most of its members, so it allows the permeabilization of knowledge across the company and, consequently, its consolidation. Coordination is another requirement in this sense. In hierarchical organizations, operations are developed in terms of formal requirements. On the opposite, organic (horizontal) structures require of a mutual adjustment and coordination among the different parts of the firm. By setting companies this way, they will become more agile and prone to changes when needed.

The second main feature of a learning organization is the development and acquisition of new knowledge. This is something that has already been explained (albeit briefly) in Chaps. 2 and 3. According to Argote and Hora (2017), learning in a firm involves the processes of creating, retaining and transferring knowledge, having this a considerable impact in the performance of an organization. Thus, learning involves a sequence of activities that move from the early identification of information needs to its final spread within the organization. Nonetheless, these activities should not be developed indiscriminately, as a key consideration must be taken into account. Learning is an activity that implies considerable costs in terms of time and resources, so it has to be directed toward the management of knowledge that is useful for the strategic goals of the firm, i.e., its successful. Of consideration should also be the election between creating knowledge internally, by means of collaboration or by market transactions, as each of them has its pros and cons. For example, the creation of knowledge within the company allows the possession of a unique resource, but it requires of a considerable effort to be developed, as well as efficient mechanisms that ensure its protection by avoiding its leak to the competitors. By establishing collaboration (for example, through the previously explained co-opetition), a firm is allowed to develop a knowledge of its interest while sharing costs and risks with other entities at the expense of sharing its possession and a true commitment with the collaborators. Finally, gathering knowledge via market transactions ensures a rapid and accessible source to certain information that may be of interest to the company. However, that information is usually available to other firms (competitors) and is of generic nature in most of the cases, meaning by this that it is not specifically developed for the firm that does purchase it. Even in some context, the combination of different approaches is recommended (Ferraris et al. 2017) or even necessary (Grigoriou and Rothaermel 2016).

Finally, we see some elements related to the organizational culture when de Geus (2002) says that a learning organization requires openness to the environment, avoiding punishment to mistakes and an alignment of interest with all its members. By those features, a learning organization is directed toward change by considering not only its own state, but also the reality outside its walls (openness to the environment), by allowing mistakes (to a certain extent) as an unavoidable part of the learning process, and an alignment of interests between the members of the organization and the organization itself, so that the actions developed are in accordance with the real needs of the firm.

Thus, the concept of learning organization shows how a firm should be in the context of the twenty-first century, characterized by environmental volatility uncertainty, complexity and ambiguity (VUCA), as exposed by Barber (1992), one of the early users of the acronym. Constant technological developments, geopolitical and war conflicts, economic shocks and even two pandemic events have taken place since the beginning of the 1900s. This requires firms to be constantly screening the environment to detect changes that can threaten their survival prospects, and thus the sustainability of their activities. That screening leads to the gathering of data that, by means of a learning process, becomes information that is useful to the company (this is where the need for learning organizations is demonstrated). Nonetheless, the way firms learn and, consequently, innovate toward sustainability depends on their environment, which is partly molded by the economic sector they belong to. Consequently, in the following section we will expose how a learning organization may exist in the wine industry.

4.2   The Learning Organization in the Wine Industry

The wine industry is undoubtedly linked to rural areas as their main locations. Many wineries also execute the role of vine-growers, while those which do not at least try to be close to them to ensure a proper supply of factors. Thus, we can understand that a good proportion of wineries find themselves vertically integrated and established in regions that act as sources of knowledge through factor agglomeration, the so-called clusters (Porter 1990). Consequently, businesses related to the wine sector and more specifically those that treat (directly or indirectly) with wineries must account for that reality to undertake actions toward sustainability.

Regarding the type of organizations based on the activity they develop, we find business in all primary, secondary and tertiary sectors. The business in each of them has very different goals and is related to very different elements from the environment. It is evident that vine-growers are mostly dependent on natural factors, while wineries are more affected by phenomena linked to industrial activities, such as a proper supply of water and energy, while distribution, selling, logistic and wine tourism companies are more sensitive to market requirements. In this stream of reasoning, primary sector can contribute to a greater extent to sustainability via environmental actions. Secondary sector firms can enhance their sustainability both in environmental and social actions, being the same for the tertiary sector (all three need enough economic sustainability). As all dimensions are needed for the long-term well-being of the sector, the collaboration among them is crucial. Here, learning organizations will be those that work together for the achievement of integral sustainability. In order to do so, and given the different nature of their economic activities, openness to the environment is mandatory, as the visions that each of them entails are notably different.

For that reason, the development of open innovation actions may be enriching for all parts, as it allows to know and understand everyone needs while also enriching the own ones. Thus, the binomial between commitment (with other companies) and self-awareness (of the own needs) combined with the pooling of resources from the different entities can enhance the achievement of a proper sustainability level in the sector. This support to open innovation is also enhanced by the fact that the majority of companies related to the wine industry (at least, the ones that depend exclusively on this product) are small and medium enterprises (SMEs), so that by developing co-operation actions, the resources individually needed for developing innovations are reduced compared to innovation on its own. Furthermore, the fact that wineries' and vine-growers' locations may act as a pole of attraction for wine-related organizations (whether economic or not) also helps to find other supporting actors toward the deployment of sustainability innovations.

We also have to bear in mind that current demands from consumers toward sustainability on both product and operations force wineries and related companies to guarantee those demands to keep their competitiveness (Cantele and Zardini 2018). Apart from the obvious development of environmentally friendly products (for example, organic, ecologic and natural wines), the final impact of the complete set of activities related to the product is also important. It is important to bear in mind that the value of a product is not the same as its price, and this reality determines the attractiveness for the client. Given the requirements in terms of sustainability, transparency with the final consumer becomes almost as important as its organoleptic properties. Consequently, the learning organization in the wine industry is not only able to gather these demands, but is also able to reach to the final consumer with that information. For that, formal information systems are needed, something that has been facilitated by technologies as blockchain, that allows to guarantee the practical immutability of the recorded information (Bayer et al. 1993; Chaum 1979), which saw its first implementation from Nakamoto (2008). Indeed, the application of blockchain to the traceability of the wine from the vine to the glass is of special interest, as it is a measure that allows consumers to ensure that the product they are drinking complies with their demands (what increases its perceived value) while the companies involved in the process also have a higher control on the interconnection with their adjacent companies in the supply chain, what allows for the optimization of their operations. However, the deployment of such technology requires a great number of users together with initial fixed investments, so the development of a common standard that can be used by different stakeholders in the sector is a must.

Another question related to the label of learning organization in the wine industry is its need for constant transformation (i.e., its dynamicity). It is particularly difficult in a sector that is so linked to tradition and history, where replacing them by novelty is not an option. On the contrary, learning organizations in the wine industry should have the capability to merge tradition, and more specifically ensuring the preservation of its roots, with up-to-date technologies, operations and products, so that these terms move from opposition to complementation. A good example would be combining a wine elaborated by means of traditional techniques with the latest developments in traceability (such as the mentioned blockchain) and the optimization of the label designs (for example, by using eye-tracking to determine the optimum labels depending on the client profile). Given the varied nature of such activities, the open innovation defended in the prior paragraphs is strengthened.

Regarding punishment to mistakes, it is important to say that it is a double-edged sword. By one side, certain mistakes can threaten the survival of a company, what is a necessity that cannot be ignored. On the other hand, an extreme constraint not to make any errors hinders the learning, the adaptability and, last but not least, the evolution of the company. Given this reality, the question is whether or not to allow mistakes, but which ones can be allowed. Developing a bit more this affirmation, the efforts of the company regarding errors have to be placed in avoiding those that can compromise its survival (for example, an inadequate financial strategy, or a not enough studied divestment plan), giving more freedom for other activities that, despite having the potential for hindering performance in the short term, serve to improve the knowledge of the company and, consequently, future decision-making, as in some circumstances errors enhance subsequent learning (Di Costa et al. 2018).

Finally, it is important to expose the need for the individual-organizational alignment. In the case of the wine industry, we have different stakeholders, as exposed in Chap. 3: vineyard workers, local communities, suppliers, customers, regulatory bodies, environmental groups and investors, among others. However, the need for satisfying most of their complaints often leaves on the blind spot the fact that the harmony between an organization and its members is crucial for the sustainability in all of its dimensions. We can think of a firm as a combination of different individuals that belong to a determined number of groups that have their own thoughts, beliefs and interests. These groups can be formed due to formal divisions (for example, if the company is divided according to the type of activities to develop) or by informal reasons (for example, the social and value sets of the individuals), even existing the possibility of different criteria combined. The problem here is that the firm in question will have to face many different claims within the firm. If some of those demands are divergent in interest, the different groups may act on their benefit, what would affect the company in terms of performance and survival prospects (i.e., sustainability). Focusing on the wine industry, such may be the case of a vertically integrated organization that has assumed vine-growing and the activities of distribution. In this case, the interests of the different members may be divided not only according to the different links on the supply chain, but also inside those groups, as Fig. 4.1 illustrates.

In order to cope with this reality, the participation of all the members in the formulation and implementation of strategies, regardless on the level they belong to, facilitates the solving of the problem. By considering their proposals, not only the acceptance prospects are enhanced, but also the information that is generated in each subgroup is deployed all along the organization, as it becomes explicit. By making those subgroups part of some strategic management decisions, interactions among them also occur, what helps to information deployment while simultaneously facilitating knowing the needs in other parts of the organization.

To sum up, the way a learning organization is created depends on the reality of its environment, which is partially dependent on the economic sector it is related

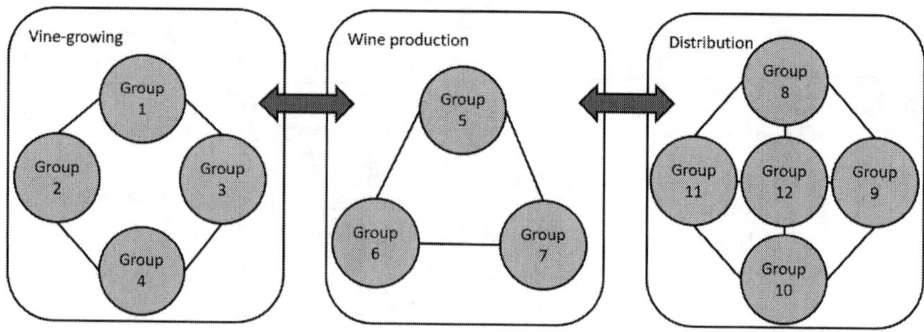

Fig. 4.1 Presence and interaction of stakeholder groups within a company. *Source* own elaboration

to. In the case of the wine industry, the prevalence of SMEs with limited resources, the need to face a VUCA environment and the unavoidable linkage to the origins of viticulture and winemaking force learning organizations to co-operation as means of innovation and knowledge management, horizontal structures that allow the spread and consolidation of information and the participation of all the collectives that are part of the firm to ensure a proper design and implementation of strategic measures, all of that being directed towards the consecution of economic, social and environmental sustainability.

References

L. Argote, M. Hora, Organizational learning and management of technology. Prod. Oper. Manag. **26**(4), 579–590 (2017)

H.F. Barber, Developing strategic leadership: The US Army war college experience. J Manage Dev **11**(6), 4–12 (1992)

D. Bayer, S. Haber, W.S. Stornetta, Improving the efficiency and reliability of digital time-stamping, in *Sequences II: Methods in Communication, Security, and Computer Science* (Springer, New York, pp. 329–334). (1993)

S. Cantele, A. Zardini, Is sustainability a competitive advantage for small businesses? An empirical analysis of possible mediators in the sustainability-financial performance relationship. J. Clean. Prod. **182**, 166–176 (2018)

D.L. Chaum, *Computer Systems established, maintained and trusted by mutually suspicious groups* (University of California, Electronics Research Laboratory, 1979)

A. De Geus, *The Living Company* (Harvard Business Press, 2002)

S. Di Costa, H. Théro, V. Chambon, P. Haggard, Q. J. Try and try again: Post-error boost of an implicit measure of agency. Exp. Psychol. **71**(7), 1584–1595 (2018)

A. Ferraris, G. Santoro, S. Bresciani, Open innovation in multinational companies' subsidiaries: the role of internal and external knowledge. Eur. J. Int. Manag. **11**(4), 452–468 (2017)

K. Grigoriou, F.T. Rothaermel, Organizing for knowledge generation: internal knowledge networks and the contingent effect of external knowledge sourcing. Strateg. Manag. J. **38**, 395–414 (2016)

N. Khatri, Consequences of power distance orientations in organisations. Vision **13**(1), 1–9 (2009)

S. Nakamoto, *Bitcoin: A Peer-to-Peer Electronic Cash System* (2008). https://bitcoin.org/bitcoin.pdf

M.E. Porter, The competitive advantage of nations. Harv. Bus. Rev. Bus. Rev. **68**(2), 73–93 (1990)

Chapter 5
Conclusion

As we have seen in the previous chapters, the wine industry is made up of a wide melting pot of organizations in terms of their nature, functions and purposes. The case of the wine industry is of particular interest for several reasons. Firstly, its strong anchoring in tradition and history, both of the product itself and of the societies in which it was produced, is a heritage that has survived countless hostilities over the centuries. Thus, from the ancient civilizations of Mesopotamia and Egypt to its current presence around the world, the origins on which the brew was founded have remained remarkably well preserved to this day. In addition, this economic activity is mostly located in rural and agricultural areas. This is of particular relevance given the role that the service sector is now taking on with the advent of new technologies. Given the opportunities presented by the latter, activities essential to human well-being seem to take a back seat, or at least for some people. As a turning point to this reality, we can present the wine industry. Strongly rooted as an industrial sector, it is closely linked to the primary sector through winegrowers, as well as to the territories where these activities are located. At the same time, it has many links with various tertiary sectors, such as distribution, sales and even tourism. Indeed, the wine sector serves as a dynamizing element for rural areas, encouraging diversification in them and acting as a pole of attraction for investment and consumers, whether in the form of wine, history and culture lovers, or simply tourists in search of new experiences.

Despite the contributions that this economic, gastronomic, leisure, historical and cultural activity can make, it is no less true that it presents a series of difficulties that it must face. On the one hand, the historical legacy on which it is based is an indispensable pillar of it that helps to the generation of added value, but it also magnifies the inertia of organizations. Like an object in free fall without friction, this historical legacy often leads to change resistance, to the necessary adaptation as the reality surrounding the organizations is altered. Since the First Industrial Revolution, it is undeniable that the context in which organizations develop has been in constant change. Thus, from the mechanization of processes to the current digitalization, technological changes have emerged not only as an aid, but as a necessity. If we

   33
E. Sánchez-García et al., *Sustainable Management Through Knowledge and Innovation*,
SpringerBriefs in Applied Sciences and Technology,
https://doi.org/10.1007/978-3-031-64792-5_5

add to this the current demands in terms of sustainability (and not only economic, but also environmental and social), we are faced with what seems to be a dilemma between maintaining tradition and responding to these demands. Fortunately, and as we have seen, these apparently opposing realities, if properly managed, can facilitate the congruence of wine organizations with the new reality and can even lead to the generation of added value with which to enhance the resilience and sustainability of the sector.

Precisely, we have been able to see how the sustainability currently defended and demanded by many is made up of these three pillars: economic, social and environmental. We have also seen that the wine industry can have a positive influence on all of them, something that, more than a possibility, is a duty for the industry. To do so, it requires the organizations that make it up to adopt an eminently innovative approach, capable of combining the latest developments with their inseparable tradition and history.

However, as is the case in most economic sectors, the wine sector is eminently made up of SMEs, so that their capacity to act in this respect is rather limited. In order to address this, we have observed how it is possible to use inter-organizational collaborations through open innovation for the development of sustainable actions, a way that, although it requires considerable coordination efforts, reduces the need for intra-organizational resources and capabilities while reducing the risks inherent to innovative activity, especially in those cases where it implies a considerable distance from the current situation. We have also seen the importance of knowledge management in this whole process. Indeed, innovation and the achievement of sustainability is not possible without adequate information to support effective and efficient decision-making. To this end, the role of stakeholders is crucial, as they are both a rich source of information as well as of power and influence, so the considerations, beliefs and goals of each of these groups cannot be ignored by the industry. Therefore, it is more than advisable to involve these stakeholders (or at least the most important ones) in the process of analysis, formulation and implementation of strategies, even more so in the case of sustainability, since it is their demands that modulate the requirements that the products and processes undertaken by the industry must meet. Consequently, information systems (both for capturing and disseminating information) become essential resources that allow, on the one hand, an adequate identification of the real needs of industry and, on the other hand, a means of communicating the actions developed in this respect. In this sense, technologies such as blockchain or the latest trends in the study of consumer behavior represent an opportunity to achieve this.

Lastly, and without wishing to extend, we would like to reflect on the learning organization as the business model to be followed in the sector. In the current VUCA environment and the paradigm shift in terms of consumer demands, companies present in the different stages of the sector's supply chain must be capable of generating, acquiring, processing and consolidating the information available to them, thereby converting it into useful knowledge for this mission. However, achieving this status inevitably requires a series of organizational transformations in a number of areas, especially in terms of design and day-to-day operations. In this sense, companies must transform the way they do business by prioritizing openness to the outside world in order to capture as much critical and useful information as possible. In addition, this information needs to permeate throughout the firm and become a powerful source of knowledge generation, which requires horizontal structures, a culture open to change and a tolerance for mistakes as part of the learning process of its members. With respect to the latter, it is also important that the different groups within the organization have an alignment of interests with it. As this is particularly difficult in the case of the wine industry, given the chain of different types of activities carried out and the diverse professional profiles involved, opting for the aforementioned option of including the most relevant groups with the greatest power of influence in the actions toward achieving learning organizations is a logic decision.

We would therefore like to conclude this brief but diverse work by emphasizing the need for a transformation in the wine sector. The strength demonstrated over the centuries by this activity transcends mere economic activity, which represents a great opportunity to improve the lives of the different social, economic and labor collectives that, both directly and indirectly, are related to it. There are not so many sectors with such a variety of dimensions linked to history as wine: we find connections with different periods, social and technological contexts and, why not say it, also human ones. The heritage it has allowed us to preserve is rich, diverse and nuanced (as is the product), so it is a great opportunity, and indeed a duty of our society, to ensure that this industry remains strong and resilient, maintaining the heritage and legacy it has bequeathed to us in return for helping in its transformation to improve the future prospects of both the sector itself and the organizations that make it up from an individual perspective, and thereby contributing to the well-being of the people who depend and will depend on this glorious brew.